Climbing Mount Everest

Understanding Commutative, Associative, and Distributive Properties

Therese Shea

PowerMath™

The Rosen Publishing Group's
PowerKids Press™
New York

Published in 2005 by The Rosen Publishing Group, Inc.
29 East 21st Street, New York, NY 10010

Book Design: Michael Tsanis

Photo Credits: Cover, pp. 5 (inset), 10 (background) © Corbis; p. 10 © Hulton-Deutsch Collection/Corbis; p. 13 © John Van Hasselt/Corbis; pp. 14, 26 © Galen Rowell/Corbis; p. 16 © Reuters New Media Inc./Corbis; p. 21 © Rob Howard/Corbis; p. 21 (inset) © Hulton Archive/Getty; p. 22 © Martin Schoeller/Corbis; p. 25 © The Image Bank; p. 29 © Robert Holmes/Corbis; p. 30 © David Keaton/Corbis.

Library of Congress Cataloging-in-Publication Data

Shea, Therese.
    Climbing Mount Everest : understanding commutative, associative, and distributive properties / Therese Shea.
        p. cm. — (PowerMath)
    Includes index.
    ISBN 1-4042-2939-6 (lib. bdg.)
    ISBN 1-4042-5142-1 (pbk.)
    6-pack ISBN 1-4042-5143-X
    1. Commutative law (Mathematics)—Juvenile literature. 2. Associative law (Mathematics)—Juvenile literature. 3. Distributive law (Mathematics)—Juvenile literature. 4. Everest, Mount (China and Nepal)—Juvenile literature. I. Title. II. Series.
    QA115.S53 2005
    512—dc22
                                    2004005317

Manufactured in the United States of America

Contents

Natural Wonder of the World   4

Man Against Nature   11

Preparing for the Climb   18

From Base Camp to Summit   24

"Because It's There"   30

Glossary   31

Index   32

Natural Wonder of the World

The people of Tibet call it Chomolungma (choh-moh-LUNG-muh), which means "Goddess Mother of the Earth." Those in Nepal call it Sagarmatha (SAH-quhr-MAH-tuh), meaning "Goddess of the Sky." This mountain, located on the border of Tibet and Nepal in the Himalaya mountain system of southern Asia, was once known to English speakers as Peak XV (15). Today, it is better known as Mount Everest.

British **surveyors** measured the summits, or highest points, of the Himalayas between 1847 and 1854. The tallest peak in this mountain system was found to be the tallest peak of any mountain ever measured. In 1865, the chief surveyor named this mountain after the previous surveyor general of India, Sir George Everest. The official height of Mount Everest is still debated, although most believe it is between 29,028 feet and 29,141 feet. This is about $5\frac{1}{2}$ miles above sea level.

Many people from all over the world hoped to be the first to reach the top of Mount Everest. Even after that was accomplished, people continued to think of the climb to the top of Mount Everest as the greatest **feat** of human ability. The climb takes strength, courage, intelligence, and especially luck.

The Himalaya mountain system is about 1,500 miles long (2,414 kilometers). It runs through several countries, including Nepal, Bhutan, and India. It also runs through Tibet, a region of China.

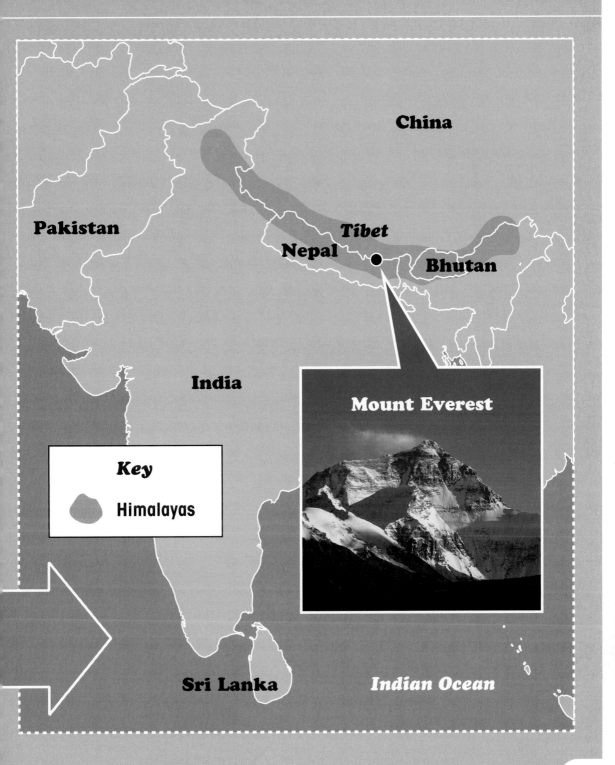

China

Pakistan

Tibet

Nepal

Bhutan

India

Mount Everest

Key

Himalayas

Sri Lanka

Indian Ocean

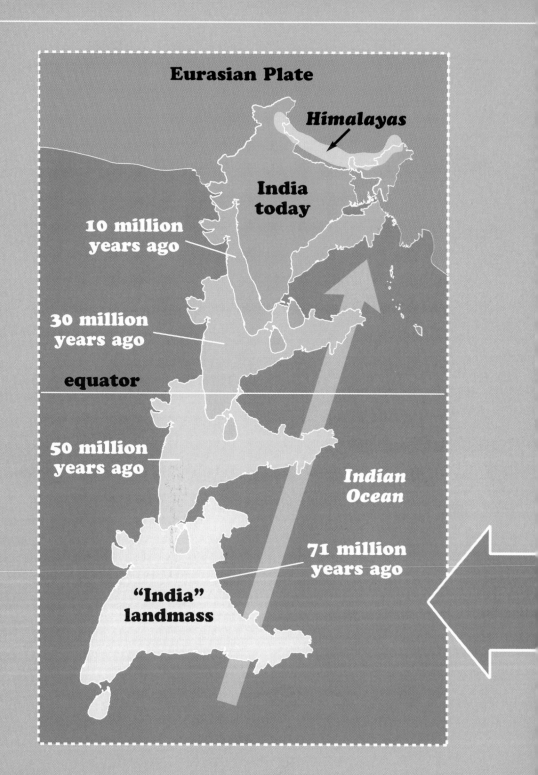

Eurasian Plate

Himalayas

India today

10 million years ago

30 million years ago

equator

50 million years ago

Indian Ocean

71 million years ago

"India" landmass

Like the Grand Canyon and Niagara Falls, Mount Everest is considered a natural wonder of the world. Every year, people from all over the world travel to see its height and beauty. Studying how Everest formed teaches us how Earth came to look as it does today.

Geologists believe that Earth's upper surface rests upon deep, moving slabs called plates. These plates slide very slowly in different directions. When 2 plates run into or move away from each other, earthquakes and volcanoes occur. The force of 2 plates pushing against each other pushes Earth's surface upward to form mountains. This theory of plate movement is known as plate tectonics (tek-TAH-niks). Today, geologists believe that there are 13 major plates still moving and changing Earth's surface.

Millions of years ago, the plate carrying an island that became India collided with the plate carrying Asia. One plate pushed beneath the other, forcing it upward. The result was the huge rock formations that are now the Himalaya mountains. The tallest of these formations is Mount Everest. It is believed that these plates are still pushing against each other, forcing Mount Everest to grow taller every year!

The map on page 6 shows the movement of the India landmass over millions of years. Seventy-one million years ago, it was south of the equator. The India landmass slowly moved north until it collided with the plate carrying the landmass of Asia and Europe.

We can use a basic **property** of math to figure out how high Mount Everest may be in the future. The **commutative property** of addition states that when the order of 2 or more **addends** is changed, the sum will stay the same. Let's say that Everest rose 5 millimeters in 1 year and 18 millimeters the next year. How many millimeters did it rise altogether? To find the answer, we need to add the 2 measurements. Let's see what happens when we apply the commutative property to this problem.

---

commutative property of addition:
$$a + b = b + a$$

5 millimeters + 18 millimeters = 18 millimeters + 5 millimeters

$$
\begin{array}{r}
5 \text{ millimeters} \\
+ 18 \text{ millimeters} \\
\hline
23 \text{ millimeters}
\end{array}
\quad = \quad
\begin{array}{r}
18 \text{ millimeters} \\
+ 5 \text{ millimeters} \\
\hline
23 \text{ millimeters}
\end{array}
$$

Mount Everest rose 23 millimeters in 2 years.

Let's try an example demonstrating the commutative property of multiplication. This property states that when the order of 2 or more factors is changed, the product stays the same. If Mount Everest rose 1.5 millimeters every year, how many millimeters would it have risen in 25 years? To figure out this problem, you need to multiply 1.5 millimeters by 25 years.

commutative property of multiplication:
a x b = b x a

1.5 millimeters per year x 25 years = 25 years x 1.5 millimeters per year

$$
\begin{array}{r}
1.5 \text{ millimeters per year} \\
\times\ 2\ 5 \text{ years} \\
\hline
7\ 5 \\
+\ 30 \\
\hline
37.5 \text{ millimeters}
\end{array}
$$

=

$$
\begin{array}{r}
2\ 5 \text{ years} \\
\times\ 1.5 \text{ millimeters per year} \\
\hline
12\ 5 \\
+\ 25 \\
\hline
37.5 \text{ millimeters}
\end{array}
$$

Mount Everest would have grown 37.5 millimeters in 25 years.

Some scientists estimate that Earth's plates move between 1 centimeter and 10.2 centimeters a year. Let's say that 1 plate is moving 2.5 centimeters a year. How far would the plate move in 75 years? Remember, the order of the numbers does not change the product.

2.5 centimeters per year x 75 years = 75 years x 2.5 centimeters per year

$$
\begin{array}{r}
2.5 \text{ centimeters per year} \\
\times\ \ 7\ 5 \text{ years} \\
\hline
12\ 5 \\
+\ 175 \\
\hline
187.5 \text{ centimeters}
\end{array}
$$

=

$$
\begin{array}{r}
7\ 5 \text{ years} \\
\times\ \ 2.5 \text{ centimeters per year} \\
\hline
37\ 5 \\
+\ 150 \\
\hline
187.5 \text{ centimeters}
\end{array}
$$

The plate will move 187.5 centimeters over 75 years.

From the moment the British survey team announced that Mount Everest was the highest mountain in the world, people began dreaming of being the first to stand on its summit. Many people became famous for their attempts. Among the most famous of these early explorers of Mount Everest was George Mallory. Mallory tried to **scale** the mountain 3 times. Each time, Mallory and the men in his **expedition** risked their lives in dangerous situations.

In June 1924, Mallory made his third attempt up the north side of the mountain with a man named Andrew "Sandy" Irvine. They were last seen near the base of the summit, less than 1,000 feet from the top. This is the last time that Mallory and Irvine were seen alive. It is still unknown whether or not they made it to the top of the mountain. Mallory's body was found 75 years later, when a special group was formed in 1999 to recover the bodies of the 2 men. Why hadn't anyone attempted this before? Conditions are so harsh on Everest that every bit of energy must be saved for the downward journey. One climber estimated that it would take the strength of 6 men to carry 1 person down the mountain.

George Mallory (second from the left in the back row) and Andrew Irvine (far left in the back row) are shown here with other mountain climbers during the British expedition up to Mount Everest in 1924.

One obstacle that climbers of Mount Everest face is wind. On an ideal day for climbing, the wind on Everest's summit can measure about 40 miles per hour. Between November and February, the wind can reach more than 180 miles per hour. That is more than twice as strong as hurricane winds. These strong winds can easily sweep a person off the icy mountain.

Climbers ascend from **base camp** to a series of 4 other camps as they make their way along one route traveling southeast from the Nepal side. Let's say that a climber measures the wind at base camp at 40 miles per hour. He radios to the next highest camp and is told that the wind is 20 miles per hour faster there. Climbers at the top say the wind increased an additional 50 miles. How strong is the wind at the summit? We need to add these numbers. The **associative property** of addition states that when the grouping of addends is changed, the sum stays the same. The parentheses are used to indicate the grouping of numbers.

associative property of addition:
$$(a + b) + c = a + (b + c)$$

$$(40 + 20) + 50 = 40 + (20 + 50)$$

40 miles per hour		20 miles per hour
+ 20 miles per hour		+ 50 miles per hour
60	=	70
+ 50		+ 40
110 miles per hour		110 miles per hour

The wind at the summit is 110 miles per hour.

Extreme temperatures are another danger. In January, the coldest month, the temperature at the summit averages about −33°F (−36° Celsius). In July, the warmest month, the average summit temperature is about −2°F (−19° Celsius). Even with proper clothing, climbers often get **frostbite**, which can cause a person to lose feeling in their fingers and toes. Climbers can also suffer from **hypothermia**, a condition in which the body temperature drops below normal, affecting the working of the muscles and the brain. Gradually, body functions stop.

Another challenge of climbing Everest is a lack of oxygen. The higher the elevation, the less oxygen is available for a person's lungs to take in. Most people begin to feel sleepy or weak and lose the ability to think clearly. Today, many mountain climbers use oxygen tanks, like those pictured here.

Three climbers plan to make a trip to the summit. Each climber plans to use 2 liters of oxygen per minute for 2 hours. How many liters of oxygen do the climbers need altogether? We can use the associative property of multiplication to figure out this problem. This property states that when the grouping of numbers is changed when multiplying, the product will stay the same.

associative property
of multiplication:
$$(a \times b) \times c = a \times (b \times c)$$

3 climbers
2 liters per minute
2 hours of oxygen needed = 120 minutes

$$(3 \times 2) \times 120 = 3 \times (2 \times 120)$$

3 climbers		2 liters per minute
x 2 liters per minute		x 120 minutes
6	=	240
x 120 minutes		x 3 climbers
720 liters		720 liters

The 3 climbers need
720 liters of oxygen.

After Mallory and Irvine disappeared in 1924, other expeditions continued to try to reach the top of Everest but failed due to poor weather conditions and inferior clothing and equipment. In 1953, Britain sent its tenth expedition to the south side of Everest with oxygen tanks, warm clothing, and up-to-date climbing equipment. They set up a series of camps on the mountain to allow their bodies to gradually adjust to the low-oxygen atmosphere, which is a system still practiced today. Two men scaled the mountain first, but had to turn back due to a broken oxygen tank when they were only about 300 feet from the top.

The group sent 2 more men up the mountain. One was a beekeeper from New Zealand named Edmund Hillary. The other man, Tenzing Norgay, was a Sherpa (SHUR-puh), a member of a Tibetan people who live mostly in the high altitudes of the Himalayas in eastern Nepal. The pair left their camp at 27,900 feet and slowly walked along a narrow 400-foot-long ridge of rock and snow called the Cornice Traverse. To the right of the ridge is a drop of 10,000 feet. An 8,000-foot drop lies to the left.

After passing the ridge, Hillary and Tenzing Norgay came to a 40-foot "step" of snow and ice. Hillary found a crack in the step. He wedged his body in the crack and worked his way up. Today, climbers use ropes and hooks on what is now known as the Hillary Step. Tenzing and Hillary had passed the last obstacle. On May 29, 1953, they finally reached the summit of Mount Everest.

Since Tenzing Norgay and Edmund Hillary scaled Everest in 1953, safety equipment has improved. This picture of a guide climbing an icy cliff on Everest shows his ice axes, boot spikes, and rope tied around his waist.

This picture of Tenzing Norgay (left) and Edmund Hillary (right) was taken in June 1953 in Nepal, shortly after they completed their climb up Mount Everest.

The partnership and friendship of Edmund Hillary and Tenzing Norgay began a long relationship between climbers and the Sherpa people. Because Sherpas live in the high altitudes of the Himalayas, their bodies have adapted to the lower oxygen levels in the atmosphere. They have fewer physical problems on Everest than the average climber. They can carry more, walk faster, and use fewer oxygen tanks. Therefore, expeditions often hire Sherpas to find the best routes up the mountain and carry supplies for them. These jobs are very dangerous. One-third of all deaths on Mount Everest are Sherpas. Many climbers owe their lives and their success in climbing Everest to the Sherpas.

About 5,000 Sherpas live in the Khumbu Valley, the entrance to the southern side of Everest. If 20,000 live elsewhere in Nepal, and 10,000 live in nearby countries, how many Sherpas are there altogether in this area? To figure this out, you can use the associative property of addition and add the numbers in any grouping.

5,000 Sherpas in the Khumbu Valley
20,000 Sherpas in other parts of Nepal
10,000 Sherpas in nearby countries

5,000		10,000		20,000
+ 20,000		+ 20,000		+ 5,000
25,000	=	30,000	=	25,000
+ 10,000		+ 5,000		+ 10,000
35,000		35,000		35,000

There are 35,000 Sherpas altogether.

Preparing for the Climb

Now that you have read about a successful expedition up Mount Everest, perhaps you are interested in climbing to the summit. First, let's figure the cost of supplies using the table below.

As you can see, this adventure is going to cost a lot of money! For instance, if you need to buy an oxygen mask and regulator plus a set of 4 oxygen bottles each for you and a friend, what is the total cost? To find the answer to this problem, we can use another property of mathematics called the **distributive property** of multiplication over addition. The distributive property tells us how to multiply a single term and 2 or more terms inside parentheses. Page 19 shows two ways to solve our problem.

Some guides tell climbers to plan on spending about $65,000 to climb Mount Everest! This table shows a partial list of the items a person will need for the climb. The cost per item is an estimate.

item needed	cost
airline ticket to Kathmandu, Nepal	$800
oxygen mask and regulator	$400
4 oxygen bottles (4 liters)	$1,760
1 yak	$50
shipping tents, food, etc.	$250
climbing permit in Nepal	$25,000
clothing and climbing gear	$8,000
tent	$300
food	$1,200
permit for walkie-talkies	$50

distributive property of multiplication over addition:
$$a \times (b + c) = (a \times b) + (a \times c)$$

2 x ($1,760 + $400)

Because $1,760 + $400 is in parentheses, you have 2 choices:

1. You can add $1,760 + $400, then multiply it by 2. This will give you an answer of $4,320.

$$\begin{array}{r} \$1,760 \\ +\ \ 400 \\ \hline \$2,160 \end{array} \qquad \begin{array}{r} \$2,160 \\ \times\ \ \ 2 \\ \hline \$4,320 \end{array}$$

2. Another way is to multiply each of the numbers in the parentheses by 2, then add the two products. Either way you solve the problem, your answer is $4,320.

$$\begin{array}{r} \$400 \\ \times\ \ 2 \\ \hline \$800 \end{array} \qquad \begin{array}{r} \$1,760 \\ \times\ \ \ \ 2 \\ \hline \$3,520 \end{array} \qquad \begin{array}{r} \$3,520 \\ +\ \ 800 \\ \hline \$4,320 \end{array}$$

Locate the cost of clothing and equipment in the table on page 18. This includes items such as long underwear, hiking boots, down jacket, wind pants, double boots, wool hat, crampons (boot spikes), face mask, heavy gloves, down suit, ski goggles, water bottles, ropes, and hooks. The actual list is much longer than this. All these items help protect you from the dangers of wind, temperature, and unstable surfaces. It can be especially difficult to pack for Mount Everest because of the severe temperature changes. While hiking to base camp, the sunlight reflecting off the snow and ice may make you feel very warm even though it is only 45°F. At the summit it could be 0°F.

How much will you and your friend spend on clothing, equipment, and oxygen masks? Remember, the distributive property shows us 2 ways of solving this problem. Both methods give us the same answer.

(cost of clothing and climbing gear + oxygen mask) x 2 people

($8,000 + $400) x 2 = total cost

$8,000		$8,400
+ 400		x 2
$ 8,400		$16,800

The total cost will be $16,800.

(2 x cost of clothing and climbing gear) + (2 x cost of oxygen mask)

$8,000	$400	$16,000
x 2	x 2	+ 800
$16,000	$800	$16,800

The total cost will be $16,800.

Climbers during George Mallory's time relied on wool and cotton clothing. Today's climbers dress in specially made materials that help keep them warm and dry.

Psang Tendi (far right), shown here with his wife and grandchild, is a Sherpa who worked on several expeditions to the top of Mount Everest.

Now that you have collected some of your supplies, it's time to start looking for a team to lead you up the mountain. Unless you are an expert climber, you need other people to help you find the best path and to help you if you're injured. Even expert climbers like George Mallory and Edmund Hillary, who had been on multiple expeditions to Everest, had difficulties on the mountain. The more people you have to assist you, the greater your chance of reaching the summit.

Here's a table of some of the people involved in an expedition and the fees they may charge. Each person has a special role to play. For example, the liaison (LEE-uh-zahn) officer makes sure that everyone on the expedition follows local laws and regulations.

expedition member	cost
lead guide	$25,000
assistant guide	$10,000
Sherpa	$5,000
cook	$3,500
liaison officer	$3,000
doctor	$4,000

Imagine that 3 Sherpas are needed on your expedition. How much money is needed to hire the Sherpas?

$5,000 per Sherpa x 3 Sherpas needed

commutative property of multiplication

$5,000 x 3 = 3 x $5,000

$$\begin{array}{r} \$5{,}000 \\ \times\ 3 \\ \hline \$15{,}000 \end{array} \qquad \begin{array}{r} 3 \\ \times\ \$5{,}000 \\ \hline \$15{,}000 \end{array}$$

You would need $15,000 to pay the 3 Sherpas.

From Base Camp to Summit

By now you have selected a crew to get you safely to the top. Your lead guide may show you 2 routes to the top, known as the North Route and the South Route. The North Route—up the north side of Everest—is the route that Mallory and Irvine took, and is considered the more dangerous route. The South Route, which was taken by Hillary and Tenzing, is the most popular route to the summit of Mount Everest. Which route would you choose?

Hopefully, as a new climber, you chose the South Route. At the start of the South Route, you will spend a few days at base camp. Base camp is located at 17,500 feet above sea level. How will this affect you? Climbing the mountain too fast can cause headaches and illnesses. Above 10,000 feet, it is recommended that you ascend at a rate of only 2,000 feet per day. If each day you climbed 726 feet in the morning, rested, then climbed 235 feet in the afternoon, how far would you climb in 12 days? To solve this problem, you will need to use the distributive property. Since you climbed the same number of feet each day, you can add the number of feet climbed each morning to the number climbed each afternoon, then multiply the sum by the number of days. Can you think of another way to solve this problem?

distributive property

(726 feet + 235 feet) x 12

```
  726 feet per morning        961 feet per day
+ 235 feet per afternoon       x  12 days
  961 feet per day             _____
                               1,922
                               9 61
                               _____
                               11,532 feet
```

You would have climbed 11,532 feet in 12 days.

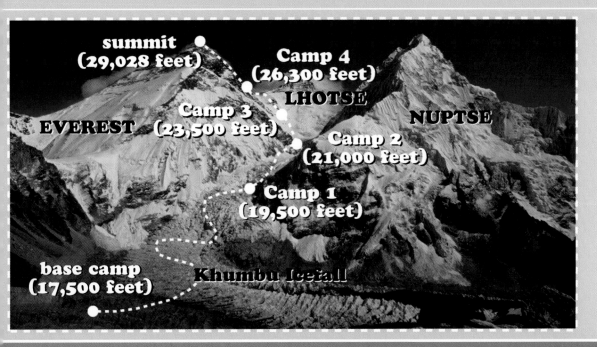

summit
(29,028 feet)

Camp 4
(26,300 feet)

LHOTSE

Camp 3
(23,500 feet)

EVEREST

NUPTSE

Camp 2
(21,000 feet)

Camp 1
(19,500 feet)

base camp
(17,500 feet)

Khumbu Icefall

If you climbed 11,532 feet from base camp, would you have climbed high enough to reach the summit? (Even though people disagree about the mountain's height, we will use 29,028 feet as the height of the summit.) The picture above shows the camps along the South Route to the top of Mount Everest.

17,500 feet—height of base camp

11,532 feet—distance ascended

commutative property of addition

$$17,500 + 11,532 = 11,532 + 17,500$$

17,500	11,532
+ 11,532	+ 17,500
29,032	29,032

The total distance climbed would be 29,032 feet.

That is 4 more feet than you need to reach the summit, which is 29,028 feet.

Many times mountain climbers have to climb at angles that measure 50 degrees or more.

The Khumbu Icefall is an expanse of pure ice. The icefall is the result of **glaciers** on the mountain slowly moving downhill. As the glaciers move down, the warmer air does not quite melt them, but causes the huge masses of ice to break into parts, like gigantic ice cubes. This is a major challenge to climbers like you. You may end up traveling this route a few times as you bring your supplies from base camp (17,500 feet) to Camp 1 (19,500 feet). Camp 2 lies further up Everest at 21,000 feet. Between Camp 3 at 23,500 feet and Camp 4 at 26,300 feet, you must pass another stretch of ice known as the Lhotse (LOHT-SAY) Face. This section is different from the Khumbu Icefall because it rises nearly 3,700 feet at steep angles. This section has a rope system that allows you to pull yourself up.

Your expedition is responsible for assembling the rope system on the risky Lhotse Face. You need to be sure you bring enough rope from Camp 3. You know the ice surface measures about 3,700 feet. You will need 2.5 feet of rope for every foot of ice. How much rope will you need?

commutative property of multiplication

$$3{,}700 \times 2.5 = 2.5 \times 3{,}700$$

```
  3,70 0 feet
x   2.5 feet of rope
  ----------
  1 850 0
+ 7 400
  ----------
  9,250.0 feet of rope
```

```
    2.5 feet of rope
x 3,70 0 feet
  ----------
     0 0
    00
   1 75
 + 7 5
  ----------
 9,250.0 feet of rope
```

You will need 9,250 feet of rope.

When you finally find yourself walking on rocks rather than on ice and snow, you are at 26,300 feet. This is the site of Camp 4 and a rocky area known as the South Col. This is the last camp climbers use on the southeast-ridge route to Mount Everest. You will rest here on your last night before you try to reach the summit.

Only 3,000 feet more and you will reach the summit! This is your last chance to change your oxygen tank and take a good look at the weather conditions. This is where many people turn around if the weather looks threatening. Remember that you want to make it to the top, but you want to make it back to base camp as well.

After climbing over 26,000 feet, climbing 3,000 more feet does not seem like much, but you must be conscious of each step. At this height, snow, wind, and low oxygen force climbers to advance very slowly. Think about how you are walking along the route of Tenzing Norgay and Edmund Hillary as you travel 400 uncertain feet along the Cornice Traverse and climb the icy 40-foot Hillary Step. They crossed these obstacles knowing that they were very close to the top. Imagine that you must reach the top in 6 hours in order to have time to return to camp before it gets dark. If you travel 8.5 feet every minute, will you have enough time to travel the 3,000 feet to the top?

6 hours = 60 minutes x 6 hours = 360 minutes

8.5 feet per minute x 360 minutes = 360 minutes x 8.5 feet per minute

commutative property of multiplication

```
  36 0  minutes              8.5 feet per minute
x  8.5  feet per minute    x  36 0 minutes
  180 0                        0 0
2 880                         51
3,060.0 feet                2 55
                            3,060.0 feet
```

You could travel 3,060 feet in 6 hours (360 minutes). That means you could reach the summit in less than 6 hours.

On Everest, the sun's reflection off the white snow can be almost blinding. Climbers must wear ski goggles that act as sunglasses as well as protection from the wind.

"Because It's There"

You've made it! You have scaled Everest and are on the top of the world. You see evidence of other climbers who have reached the summit before you, including pictures, flags, and even trash like empty oxygen canisters. How have you and more than 1,000 other people overcome so many dangers to reach this moment? Courage is needed, but problem-solving abilities—like knowing how and when to use the commutative, associative, and distributive properties to figure out math problems—are equally important. Knowing how to use these math skills helps us calculate information that could be lifesaving.

In the future, people will continue to use intelligence and bravery to reach places that seem impossible to explore, from the depths of space to the summits of mountains. Why? Perhaps for the same reason that George Mallory continued to scale Everest after defeat: "Because it's there."

addend (AA-dend) A number to be added to another number.

associative property (uh-SOH-shee-ay-tihv PRAH-purh-tee) The property which states that when the grouping of numbers is changed when adding or multiplying, the sum or product will stay the same.

base camp (BAYS CAMP) The starting camp with tents where food, equipment, medical supplies, and staff are available for climbers.

commutative property (kuh-MYOO-tah-tihv PRAH-puhr-tee) The property which states that when the order of two numbers is changed when adding or multiplying, the sum or product will stay the same.

distributive property (dih-STRIH-byoo-tihv PRAH-puhr-tee) The property which states that multiplying a sum by a number gives the same result as multiplying each addend by the number and then adding the products [a x (b + c) = a x b + a x c].

expedition (ehk-spuh-DIH-shun) A journey undertaken for a specific purpose.

feat (FEET) An act accomplished through courage or skill.

frostbite (FRAWST bite) A slight or severe freezing of a part of the body that can cause damage to the tissues of the exposed area.

glacier (GLAY-shur) A large amount of ice that moves slowly down a slope.

hypothermia (hy-poh-THUR-mee-uh) Below-normal body temperature.

property (PRAH-puhr-tee) In mathematics, a quality belonging to a group of numbers.

scale (SKAYL) To climb.

surveyor (suhr-VAY-uhr) A person who measures land.

Index

A

Asia, 4, 7

B

base camp, 12, 20, 24, 25, 27, 28

C

Chomolungma, 4
clothing, 13, 15, 18, 20
Cornice Traverse, 15, 28

E

Everest, Sir George, 4

H

Hillary, Edmund, 15, 17, 23, 24, 28
Hillary Step, 15, 28

I

India, 4, 7
Irvine, Andrew "Sandy," 11, 15, 24

K

Khumbu Icefall, 27
Khumbu Valley, 17

L

Lhotse Face, 27

M

Mallory, George, 11, 15, 23, 24, 30

N

Nepal, 4, 15, 17, 18
Norgay, Tenzing, 15, 17, 24, 28
North Route, 24

O

oxygen, 14, 15, 17, 18, 20, 28, 30

P

Peak XV (15), 4
plate tectonics, 7

S

Sagarmatha, 4
Sherpa(s), 15, 17, 23
South Col, 27
South Route, 24

T

temperature(s), 13, 20
Tibet(an), 4, 15

W

wind(s), 12, 20